ADDITION:

Choose your numbers and add them together to solve the mystery.

**Make your own MATH MYSTERIES
ADDITION**

ADDITION

Draw the cars.

Pick 4 numbers

☐ ☐ ☐ ☐
W X Y Z

Tyler has W cars, Kaiden has X cars, Jenny has Y cars, and Adam has Z cars. If they put all of their cars together, how many cars are there?

Answer

☐ + ☐ + ☐ + ☐ = ☐
W   X   Y   Z   ?

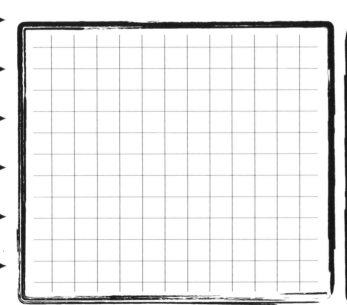

PAGE 2

Make your own **MATH MYSTERIES**

ADDITION

Draw the frogs.

Pick 3 numbers

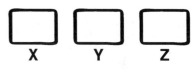

X   Y   Z

Jaymz, Ana, and Ben are catching frogs. Jaymz caught X, Ana caught Y, and Ben caught Z frogs. How many frogs did they catch?

Answer

☐ + ☐ + ☐ = ☐
X   Y   Z   ?

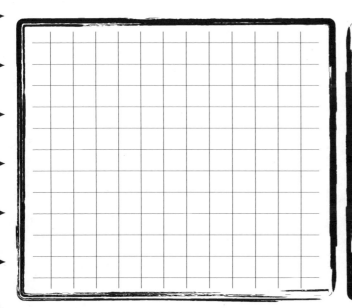

ADDITION

Draw and color Alex's suckers.

Pick 3 numbers

☐ ☐ ☐
X Y Z

Alex gave away X blue suckers, Y red ones, and Z orange suckers. How many suckers did Alex give away?

Answer

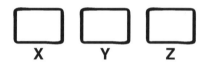

☐ + ☐ + ☐ = ☐
X   Y   Z   ?

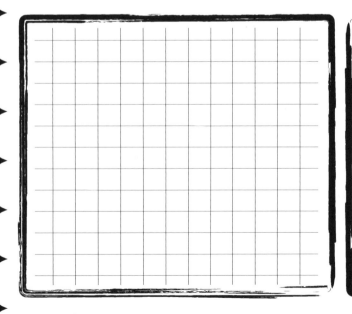

Make your own **MATH MYSTERIES**

ADDITION

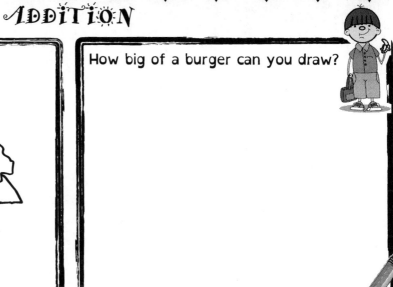

How big of a burger can you draw?

Pick 3 numbers

X   Y   Z

Adam, Daniel, and Dallas had a burger eating contest. Adam ate X burgers, Daniel ate Y burgers, and Dallas ate Z burgers. A.) Who won the first, second, and third places in the contest? B.) How many burgers did they eat all together?

Answer

☐ + ☐ + ☐ = ☐
X   Y   Z   ?

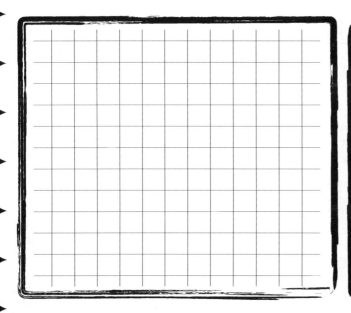

Make your own MATH MYSTERIES

ADDITION

Draw a cricket.

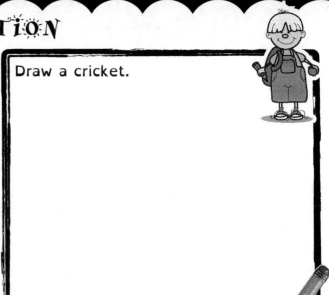

Pick 2 numbers

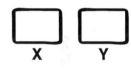

Jamel is catching crickets. He has caught X so far. How many will he have if he catches Y more?

Answer

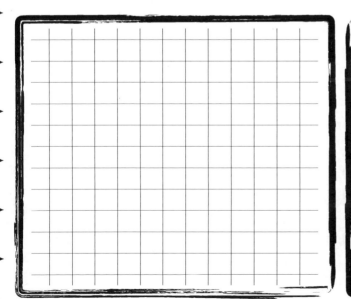

Make your own MATH MYSTERIES

ADDITION

Draw Benson's animals.

Pick 3 numbers

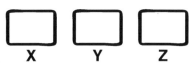
X Y Z

Benson's cat has X kittens. His dog has Y puppies. His chicken has Z chicks. How many baby animals does Benson have?

Answer

☐ + ☐ + ☐ = ☐
X   Y   Z   ?

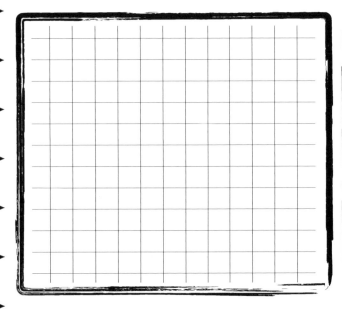

PAGE 8   Make your own MATH MYSTERIES

ADDITION

Draw Elsa's farm.

Pick 8 numbers

| A | B | C | D |
| E | F | G | H |

Elsa's job on her farm is to feed the animals. She has A cats, B dogs, C chickens, D ducks, E rabbits, F cows, G pigs, and H horses. How many animals does Elsa feed on her farm?

Answer

☐ + ☐ + ☐ + ☐ + ☐ + ☐ + ☐ + ☐ = ☐
A   B   C   D   E   F   G   H   ?

Make your own MATH MYSTERIES

ADDITION

Draw and color Josie's cake.

Pick 3 numbers

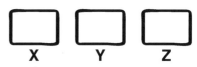

X Y Z

Josie is baking a cake. She wants to put frosting circles on the cake. Josie has X red ones, Y blue ones, and Z yellow ones. How many circles will be on Josie's cake?

Answer

$$X + Y + Z = \;?$$

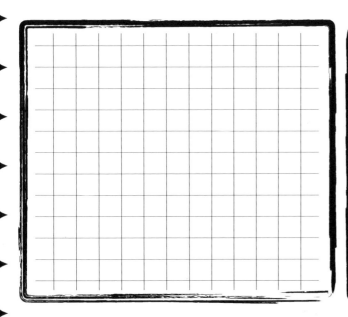

Make your own **MATH MYSTERIES**

ADDITION

Draw Anya and her cups.

Pick 4 numbers

W X Y Z

Anya has W cups of water, X cups of juice, Y cups of cocoa, and Z cups of tea. How many cups does Anya have?

Answer

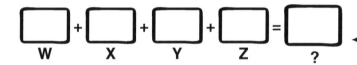

W + X + Y + Z = ?

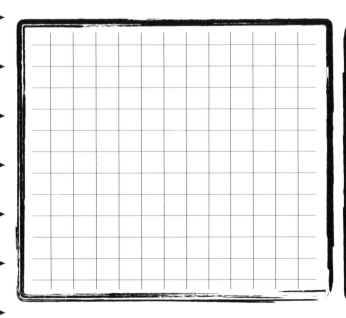

Make your own **MATH MYSTERIES**

ADDITION

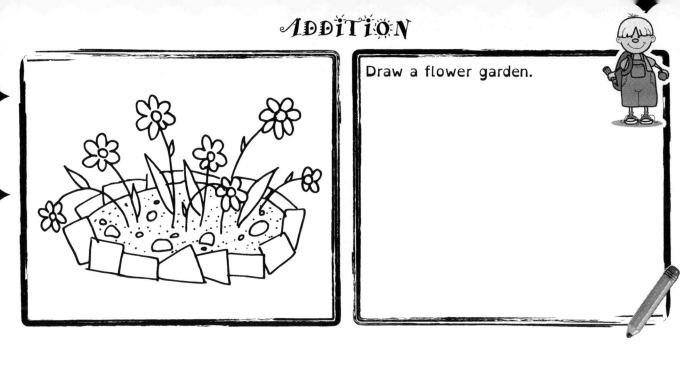

Draw a flower garden.

Pick 4 numbers

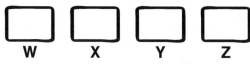
W X Y Z

Nora has a garden with W tulips, X daisies, Y orchids, and Z daffodils. How many flowers does Nora have?

Answer

$$\square_W + \square_X + \square_Y + \square_Z = \square_?$$

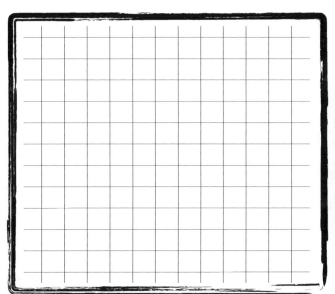

Make your own MATH MYSTERIES

ADDITION

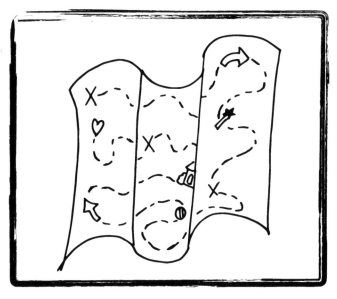

Draw a map of Miley's journey.

Pick 3 numbers

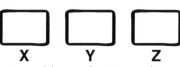

Miley wants to swim from her dock to Kacie's. First Miley must swim to a big rock, X yards away. Then she has to swim Y yards to get to a little island. After that she must swim the last Z yards to finish at Kacie's dock. How far does Miley have to swim to reach Kacie's dock?

Answer

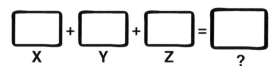

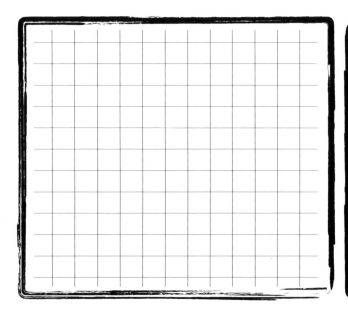

Make your own **MATH MYSTERIES**

ADDITION

Draw your own path of puddles to jump.

Pick 3 numbers

☐ ☐ ☐
X Y Z

Liam went puddle jumping. He jumped in X little puddles, Y ankle deep puddles, and Z giant puddles. How many puddles did Liam jump in?

Answer

☐ + ☐ + ☐ = ☐
X   Y   Z   ?

Make your own MATH MYSTERIES

PAGE 15

ADDITION

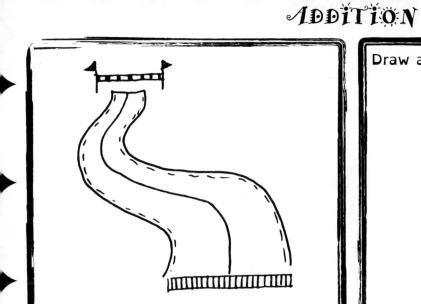

Draw a relay race track.

Pick 3 numbers

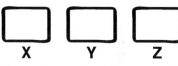
X Y Z

Ally, Randy, and Brock are running in a relay race. If Ally runs X meters, Randy runs Y meters, and Brock runs Z meters, how many meters will they run all together?

Answer

□ + □ + □ = □
X   Y   Z   ?

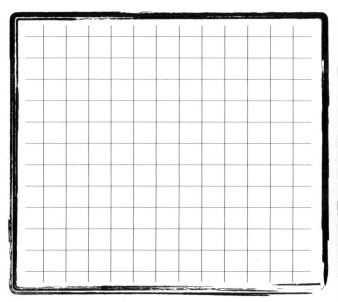

Make your own MATH MYSTERIES

ADDITION

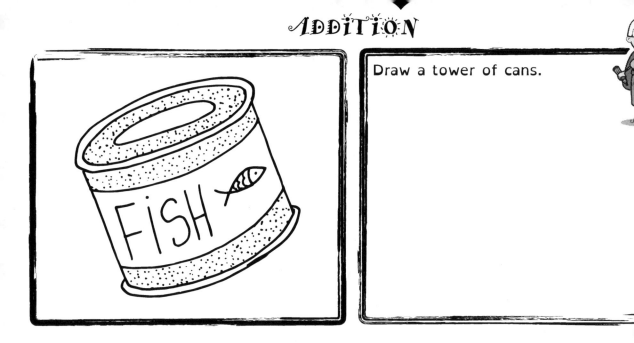

Draw a tower of cans.

Pick 3 numbers

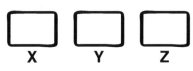
X Y Z

Cory is gathering cans to recycle. Cory found X in his house. There were Y cans in the park. He also found Z on his way to the library. How many cans has Cory collected?

Answer

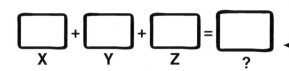
X + Y + Z = ?

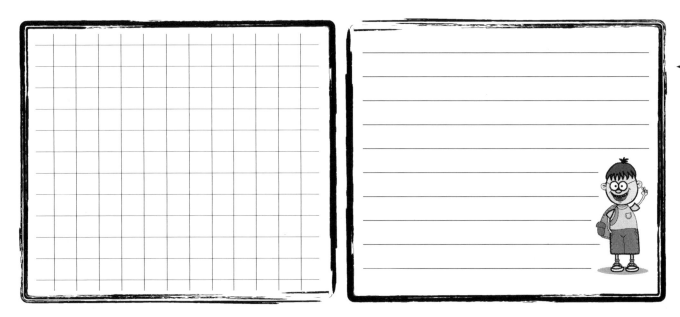

Make your own **MATH MYSTERIES**

ADDITION

Draw your favorite type of tree.

Pick 3 numbers

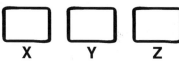
X   Y   Z

John is trying to climb to the top of a tree. There are X branches to the first resting place on the tree. There are Y branches from the middle of the tree to the second resting place. There are Z branches to get the rest of the way to the top of the tree. How many branches will John need to pass to get to the top?

Answer

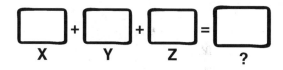
X + Y + Z = ?

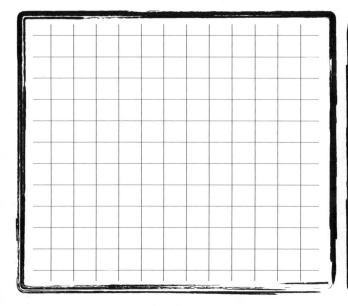

Make your own MATH MYSTERIES

SUBTRACTION:

Pick the numbers you want. But, put the biggest number first or the mystery might be hard to solve.

**Make your own MATH MYSTERIES
SUBTRACTION**

SUBTRACTION

Draw your tree house.

Pick 2 numbers
Use the larger number first

☐ ☐
X Y

You want to build a tree house. You need to have X boards. If you already have Y boards, how many more do you need?

Answer

☐ - ☐ = ☐
X   Y   ?

Make your own MATH MYSTERIES   PAGE 21

SUBTRACTION

Draw a map to your own buried treasure!

Pick 3 numbers
Make sure that the first number is at least twice as big as the others

Matthew wants to find a buried treasure X steps away. He has to take Y steps to get to the first clue. Then he needs to take Z steps to get to the second clue. How many steps does Matthew have left?

Answer

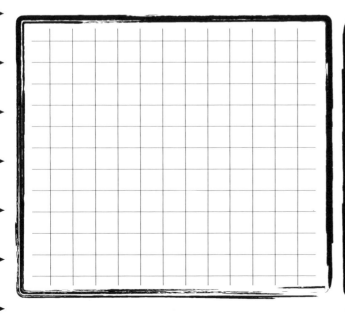

Make your own **MATH MYSTERIES**

SUBTRACTION

Draw your suitcase and toys.

Pick 2 numbers
Use the biggest number first

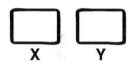

You are going on vacation and need to pack your toys. You are only allowed to bring X toys. You have packed Y toys so far. How many more toys can you pack?

Answer

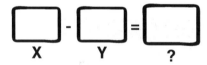

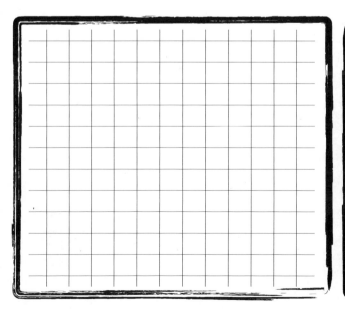

Make your own MATH MYSTERIES

SUBTRACTION

Draw a collection of pennies.

Pick 2 numbers
Use the bigger number first

X   Y

You are putting your penny collection into penny rolls. You know that you have X pennies. So far you have rolled Y pennies. How many more pennies do you need to roll?

Answer

X − Y = ?

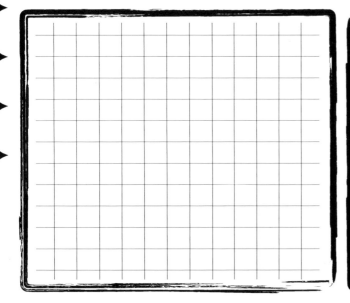

Make your own **MATH MYSTERIES**

SUBTRACTION

Draw your dream play house.

Pick 2 numbers
Use the bigger number first

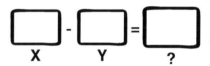
X Y

There are X people in your play house. You have Y snacks to share with them. How many more snacks do you need so that everyone gets one?

Answer

$$\square - \square = \square$$
X Y ?

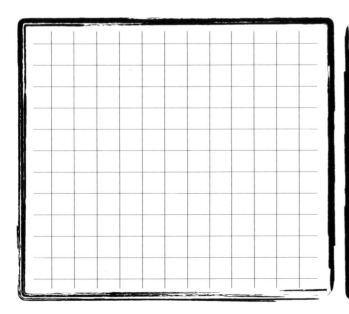

Make your own **MATH MYSTERIES**

SUBTRACTION

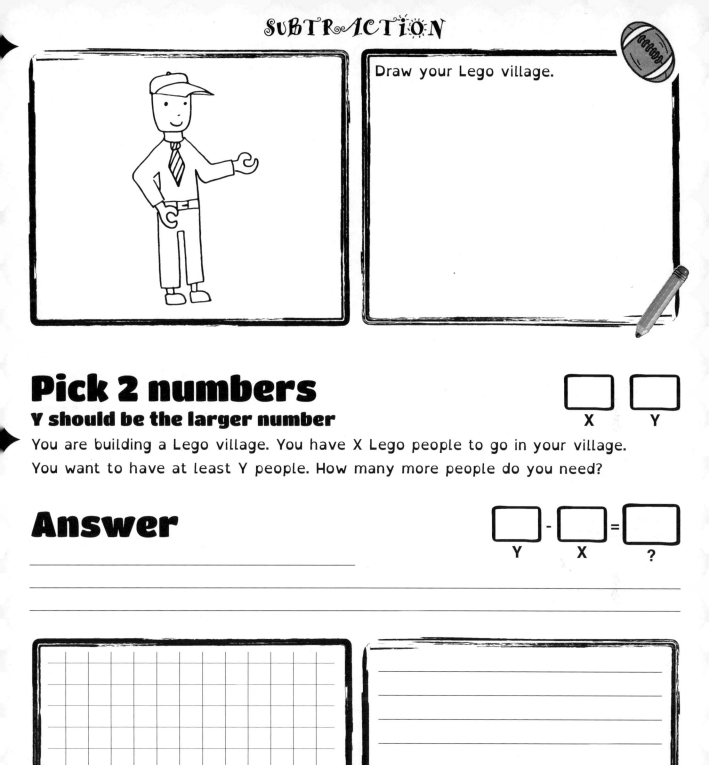

Draw your Lego village.

Pick 2 numbers
Y should be the larger number

X Y

You are building a Lego village. You have X Lego people to go in your village. You want to have at least Y people. How many more people do you need?

Answer

Y - X = ?

Make your own MATH MYSTERIES   PAGE 27

SUBTRACTION

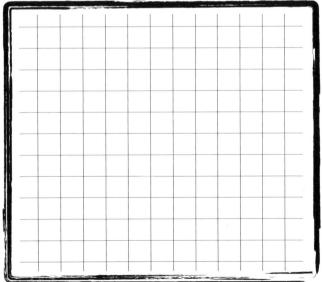

Draw a giant bath tub.

Pick 3 numbers
Make sure that the first number is at least twice as big as the others

☐ ☐ ☐
X Y Z

Betty is trying to fill her bath tub using buckets of water. It takes X buckets to fill the tub. She poured in Y buckets. Betty's little sister added Z buckets. How many more buckets of water will Betty need to fill the bath tub?

Answer

☐ − ☐ − ☐ = ☐
X   Y   Z   ?

PAGE 28   Make your own MATH MYSTERIES

SUBTRACTION

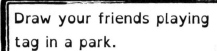

Draw your friends playing tag in a park.

Pick 2 numbers
Make X the bigger number

X Y

You want to play flashlight tag with X friends. You have Y flashlights to use. How many flashlights do you still need so that everyone has one?

Answer

$(1(\text{you}) + \boxed{}) - \boxed{} = \boxed{}$

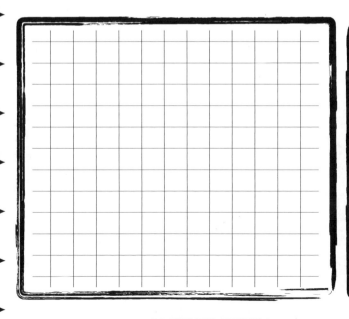

Make your own MATH MYSTERIES   PAGE 29

SUBTRACTION

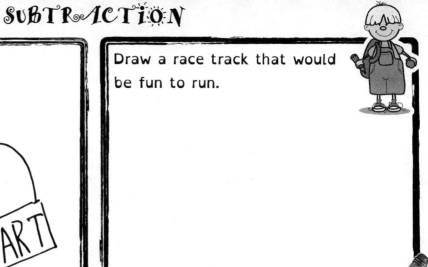

Draw a race track that would be fun to run.

Pick 2 numbers
Make X the bigger number

☐ ☐
X Y

You are running in a race. You need to run X meters to get to the end. You have already run Y meters. How much farther do you need to go?

Answer

☐ - ☐ = ☐
X   Y   ?

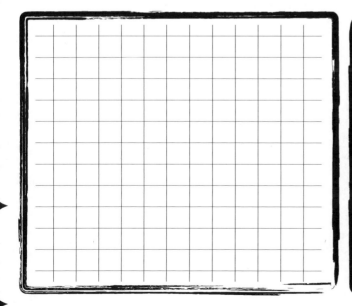

Make your own MATH MYSTERIES

SUBTRACTION

Draw a crazy cool cookie jar.

Pick 6 numbers
Make the first (U) number larger than 200 and make the rest less than 25 each

U V W X Y Z

Lilian and Brett are selling U cookies at a bake sale. They sold V cookies. Jimmy bought W cookies. Then Christa bought X cookies. Later, Rory bought Y cookies. Cory also bought z cookies. How many cookies do Lilian and Brett still have to sell?

Answer

☐ - ☐ - ☐ - ☐ - ☐ - ☐ = ☐
U   V   W   X   Y   Z   ?

PAGE 32

Make your own **MATH MYSTERIES**

SUBTRACTION

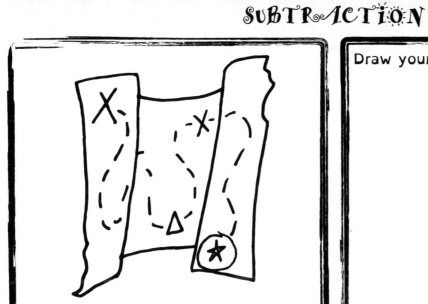

Draw your treasure map.

Pick 2 numbers
Make Y the bigger number

You have X pieces of a treasure map. There is a hidden message on the back of your pieces that say you need a total of Y pieces to complete the map. How many more pieces do you need to complete your treasure map?

Answer

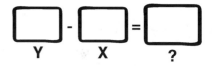

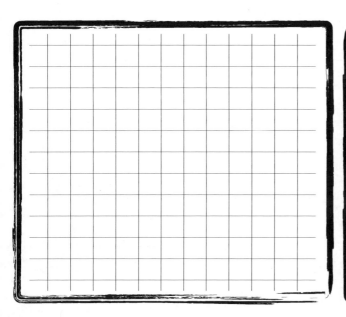

Make your own MATH MYSTERIES

SUBTRACTION

Draw a picture of something you would save your money for.

Pick 3 numbers
X should be at least twice as large as Y & Z

X Y Z

You had X quarters. You used Y quarters to buy ice cream. Then you used Z quarters to buy juice. How many quarters do you have left?

Answer

☐ - ☐ - ☐ = ☐
X   Y   Z   ?

PAGE 34   Make your own **MATH MYSTERIES**

SUBTRACTION

Draw a yard full of puddles.

Pick 3 numbers
X should be at least twice as large as Y & Z

X Y Z

Zoe and Zeke are trying to set a record for the most puddles jumped in a day. They need to jump at least X puddles. Zoe has jumped Y puddles. Zeke has jumped Z puddles. How many more puddles do they still need to jump?

Answer

X - Y - Z = ?

PAGE 36

Make your own **MATH MYSTERIES**

SUBTRACTION

Draw Brutus and his sheep.

Pick 2 numbers
X should be higher

☐ ☐
X Y

Brutus, the sheep dog, is trying to shepherd his sheep into their pin. He needs to gather all X sheep. So far, Brutus has gotten Y sheep into the pin. How many more sheep does Brutus need to shepherd into the pin?

Answer

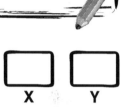

☐ - ☐ = ☐
X   Y   ?

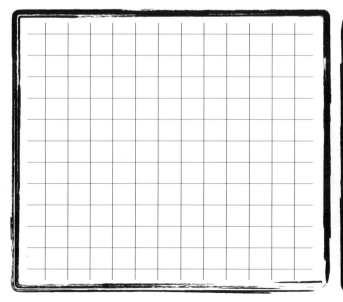

Make your own MATH MYSTERIES

PAGE 37

SUBTRACTION

Draw Peter Pickle Picker's Pickle Palace and Pickle Farm plot.

Pick 2 numbers
X should be higher

Peter Piper the pepper picker is trying to pick all the peppers on his pepper plot. There are X peppers in the plot. Peter has picked Y peppers and taken them to the pickled pepper palace. How many peppers does Peter still need to pick from the pepper plot?

Answer

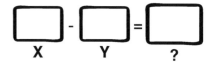

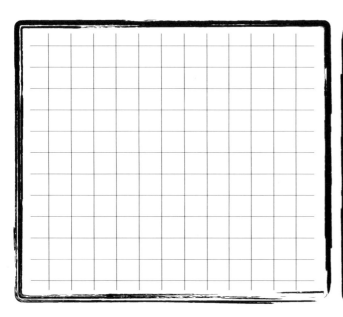

Make your own MATH MYSTERIES

MULTIPLICATION:

Choose your numbers and multiply them to solve the mystery.

**Make your own MATH MYSTERIES
MULTIPLICATION**

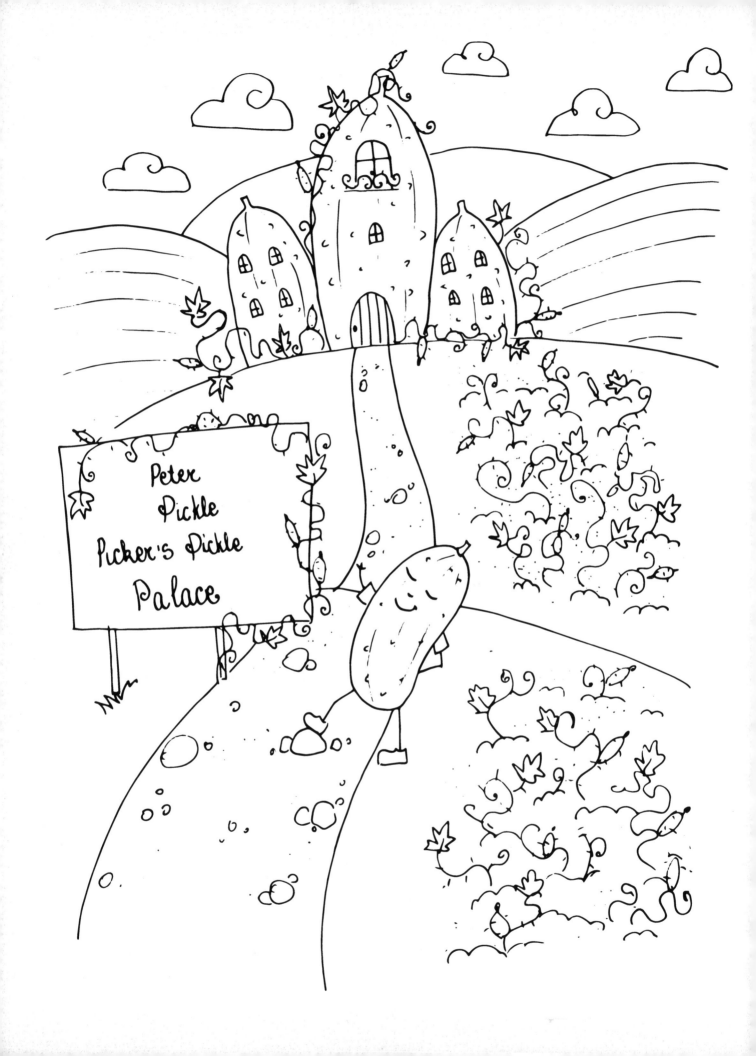

MULTIPLICATION

Draw a picture of you and your friends fishing.

Pick 3 numbers

☐ ☐ ☐
X Y Z

You and X friends want to go fishing. You each have Y fishing poles. If you need Z worms for each fishing pole, how many worms will you need to bring?

Answer

$(\square_Z \times \square_Y) \times (\square_X + 1) = \square_?$

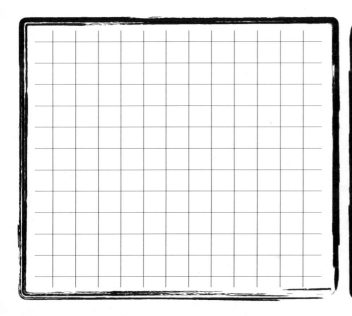

Make your own MATH MYSTERIES

MULTIPLICATION

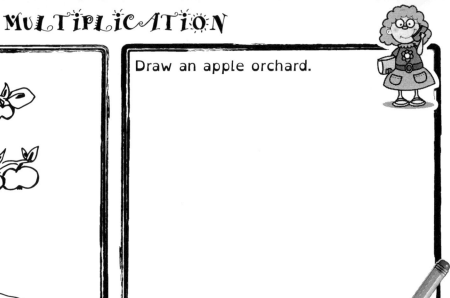

Draw an apple orchard.

Pick 3 numbers

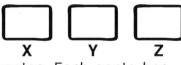
X Y Z

Joe and Laura want to plant an apple orchard. They have X apples. Each apple has enough seeds to grow Y apple trees. They want to plant Z trees.
How many more seeds do they still need?

Answer

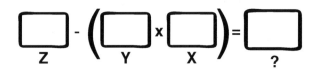

$Z - (Y \times X) = ?$

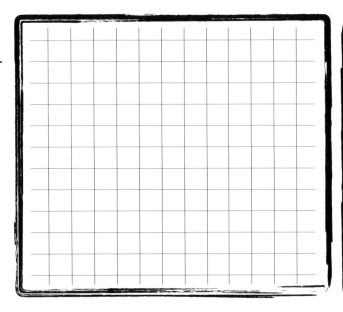

Make your own **MATH MYSTERIES**

MULTIPLICATION

Draw Jojo's birthday party.

Pick 2 numbers

☐ ☐
X Y

Jojo's birthday is coming and you want to bake cookies for the party. If X people will be at the party and everyone gets Y cookies, how many cookies do you need to bake?

Answer

☐ x ☐ = ☐
X   Y   ?

_____
_____
_____

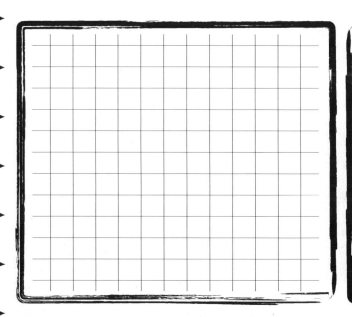

Make your own MATH MYSTERIES

MULTIPLICATION

Draw your igloo.

Pick 2 numbers

You and X friends are building an igloo. To make the igloo big enough for all of you to play inside, you will need each person to add Y blocks of ice. How many blocks will it take for you and your friends to build the igloo big enough?

Answer

$(\boxed{}_X + 1) \times \boxed{}_Y = \boxed{}_?$

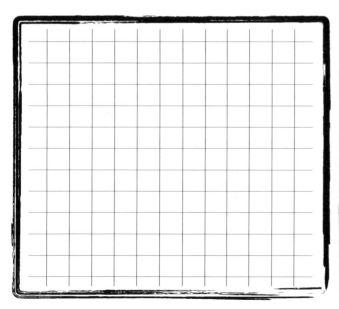

PAGE 44

Make your own MATH MYSTERIES

MULTIPLICATION

Draw Molly and her puppies.

Pick 2 numbers

Molly's dog just had X puppies! If Molly needs to give each puppy Y bottles each day, how many bottles will Molly need each day?

Answer

$$\boxed{}_X \times \boxed{}_Y = \boxed{}_?$$

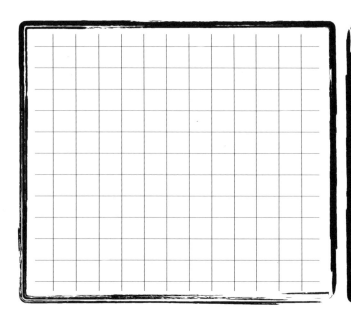

Make your own **MATH MYSTERIES**

MULTIPLICATION

Draw a park.

Pick 2 numbers

☐ ☐
X Y

Josie and Emma want to walk to the park. The park is X blocks away. Each block has Y houses. How many houses will Emma and Josie pass?

Answer

☐ x ☐ = ☐
X   Y   ?

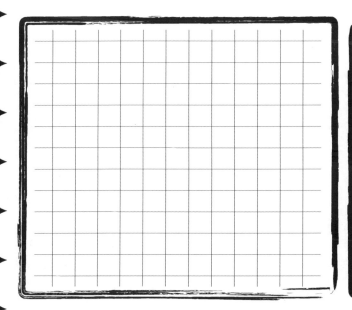

Make your own **MATH MYSTERIES** PAGE 47

MULTIPLICATION

Draw a garden.

Pick 4 numbers

☐ ☐ ☐ ☐
W X Y Z

Joe wants to plant a garden. He has W cucumbers, X zucchini, and Y onions. Each of the vegetables can grow Z plants. How many plants can Joe plant in his garden?

Answer

(☐ + ☐ + ☐) × ☐ = ☐
 W   X   Y   Z   ?

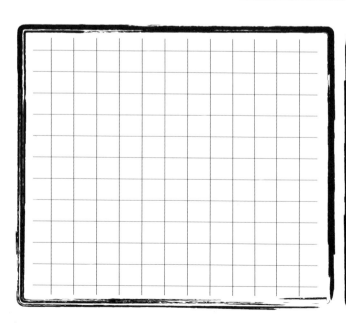

Make your own **MATH MYSTERIES**

MULTIPLICATION

What would your fort look like.

Pick 2 numbers

You want to build a fort out of boxes. You want there to be X rooms. If each room needs Y boxes, how many boxes will you need?

Answer

$$\boxed{} \times \boxed{} = \boxed{}$$
X Y ?

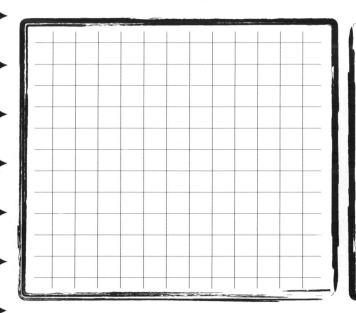

Make your own MATH MYSTERIES

MULTIPLICATION

Draw a map from your house to the store.

Pick 2 numbers

Zoey is walking to the store. It takes her X seconds to walk a yard. The store is Y yards away. How long will it take her to get to the store?

Answer

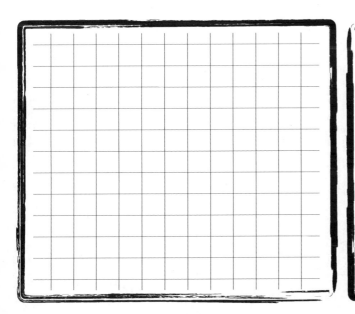

Make your own MATH MYSTERIES

MULTIPLICATION

Draw a roller coaster.

Pick 2 numbers

☐ ☐
X   Y

You are going to the theme park with X friends. You each need Y tickets so you can ride all of the rides. How many tickets will you need to buy?

Answer

(☐ +1) × ☐ = ☐
  X   Y   ?

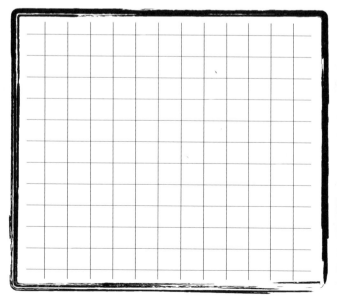

Make your own MATH MYSTERIES

MULTIPLICATION

Draw a corn field.

Pick 1 number

☐ X

Dan, Alicia, Josie, and Emma planted a corn field. They each planted enough seeds to grow X stalks of corn. If all of the stalks grow, how many corn stalks will they have?

Answer

4 × ☐ = ☐
     X   ?

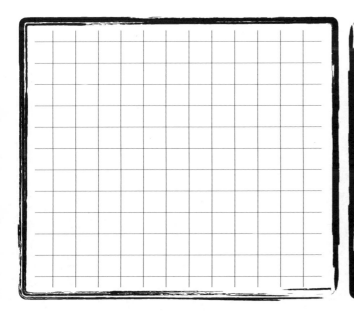

Make your own **MATH MYSTERIES**   PAGE 53

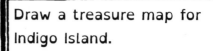

MULTIPLICATION

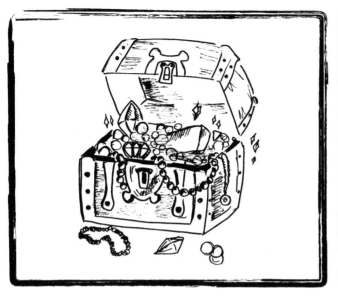

Draw a treasure map for Indigo Island.

Pick 2 numbers

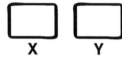

Captain Dallas and his crew (Heidi, Molly, and Roxy) are gathering their pirate treasure to bury on Indigo Island. Dallas and Heidi have X pieces of treasure each. Molly and Roxy have Y pieces of treasure each. How much treasure will they be able to bury?

Answer

$$(\boxed{}_X \times 2) + (\boxed{}_Y \times 2) = \boxed{}_?$$

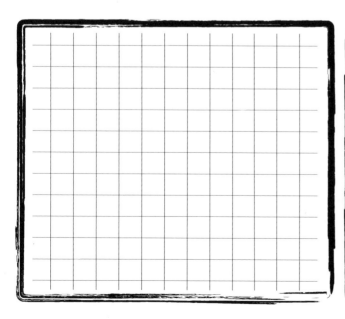

PAGE 54

Make your own MATH MYSTERIES

MULTIPLICATION

Draw their toy train.

Pick 3 numbers

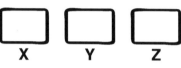
X Y Z

Adam, Jenny, Kaiden, and Tyler are building a toy train. Adam made X train cars. Jenny made twice as many as Adam. Tyler made Y train cars. Kaiden made Z times as many cars as Tyler. How many train cars do they have?

Answer ☐_X + (☐_X × 2) + ☐_Y + (☐_Y × ☐_Z) = ☐_?

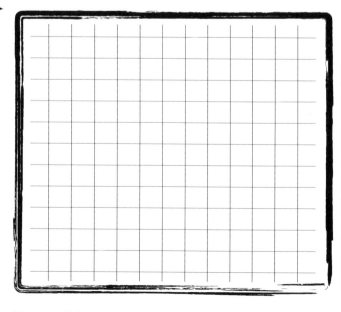

Make your own **MATH MYSTERIES**

MULTIPLICATION

Draw your new library.

Pick 2 numbers

☐ ☐
X Y

Becca, Abby, Doris, Ken, Hunter, John, and JR each have X books. Lily has Y times as many books as Abby. They gave all of their books to you to fill a library. How many books did they give you?

Answer

_____

$(7 \times \square_X) + (\square_X \times \square_Y) = \square_?$

_____
_____

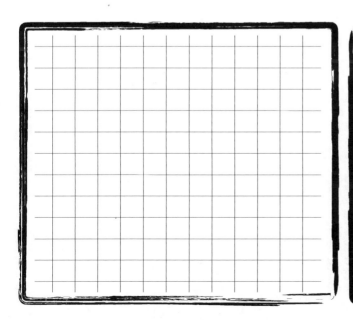

Make your own MATH MYSTERIES

MULTIPLICATION

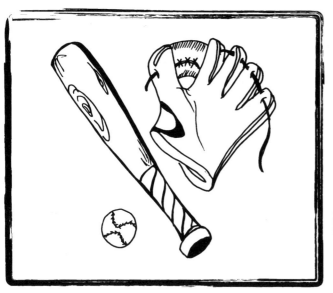

Draw yourself playing catch with your friends.

Pick 1 number

☐
X

Ethan, Emily, and Nathan are playing baseball. They each want to toss X good throws. If they can do so, how many good throws will they have tossed?

Answer

3 × ☐ = ☐
    X   ?

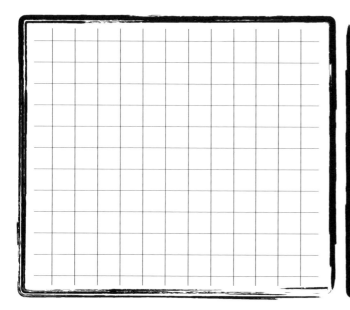

Make your own MATH MYSTERIES

DIVISION:

Pick the numbers you want to use. Sometimes you will get fractions. That is ok.

Make your own MATH MYSTERIES
DIVISION/Fractions

DIVISION/Fractions

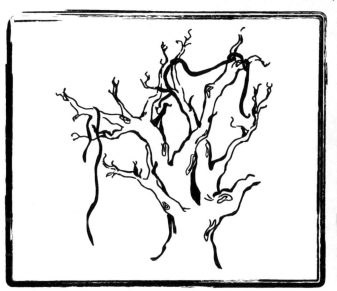

Draw a tree that you would like to climb.

Pick 1 number

□
X

Harlow is climbing a tree. The tree is X feet tall. If she is half way up the tree, how many more feet does she need to climb to reach the top?

Answer

_____

$\square_X \div 2 = \square_?$

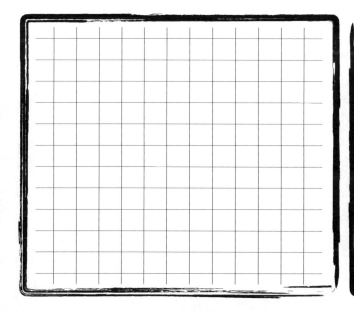

Make your own MATH MYSTERIES

DiVISION/Fractions

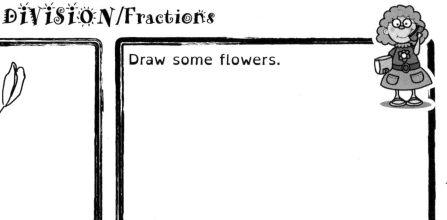

Draw some flowers.

Pick 2 numbers

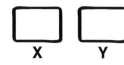
X Y

Tiffani and Vanessa are picking flowers. Tiffani has picked X flowers. Vanessa collected Y flowers. But, when they put the flowers together, their puppy ate half of Vanessa's flowers. How many flowers do they have left?

Answer

$$\boxed{}_X + (\boxed{}_Y \div 2) = \boxed{}_?$$

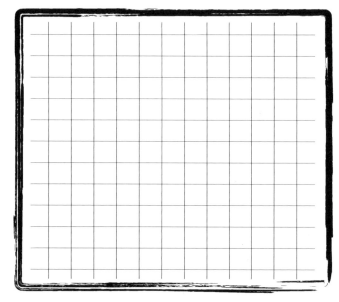

Make your own MATH MYSTERIES

DIVISION/Fractions

Draw your own tea party.

Pick 1 number

Alicia wants to have a tea party. If she has X cookies and gives each person 2 cookies, how many people can Alicia invite to her tea party?

Answer

$$X \div 2 = ?$$

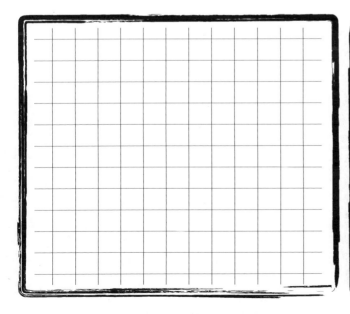

Make your own MATH MYSTERIES   PAGE 63

DIVISION/Fractions

Draw your dolls.

Pick 2 numbers

☐ ☐
X Y

Josie and Emma are doing a concert for their dolls. Emma brings X dolls and Josie brings Y dolls. Then the dog came and took one third (1/3) of the dolls away. How many dolls did the dog take from the concert?

Answer

$(\boxed{}_Y + \boxed{}_X) \div 3 = \boxed{}_?$

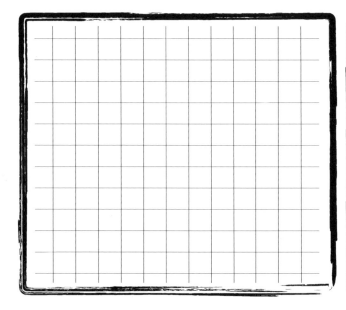

PAGE 64

Make your own MATH MYSTERIES

DIVISION/Fractions

Draw Billy and his goat.

Pick 2 numbers

X Y

Josh's goat ate his hat so he needs to get a new one. He gets paid X cents each time he milks the goat. His new hat will cost Y cents. How many times will Josh need to milk the goat?

Answer

$$\boxed{} \div \boxed{} = \boxed{}$$
Y   X   ?

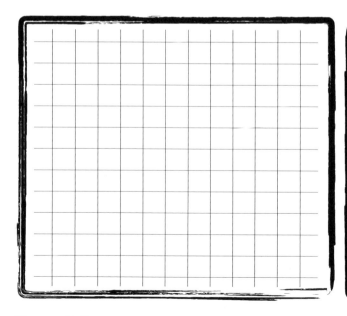

Make your own MATH MYSTERIES

DIVISION/Fractions

Draw ice cream with as many scoops as you would want.

Pick 2 numbers

☐ ☐
X Y

Angela has X pennies and wants to buy ice cream. Each scoop costs Y pennies. How many scoops can Angela buy?

Answer

☐ ÷ ☐ = ☐
X   Y   ?

Make your own MATH MYSTERIES   Page 67

DiVISION/Fractions

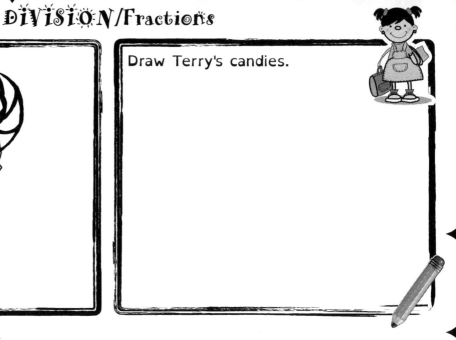

Draw Terry's candies.

Pick 1 number

☐ X

Terry has X candies. If he gives half of them to Josie, how many does he have left?

Answer

☐ ÷ 2 = ☐
X   ?

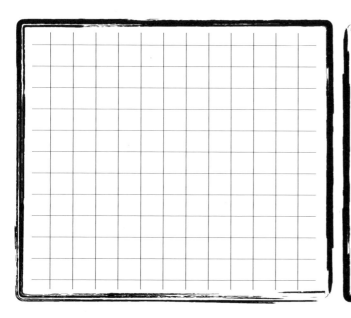

Make your own **MATH MYSTERIES**

DIVISION/Fractions

Draw a tree house.

Pick 2 numbers
The second number should be between 1 and 10

X Y

You have a tree house that is X feet high. To reach it, you need to build a step every Y feet. How many steps will you need to build to reach your tree house?

Answer

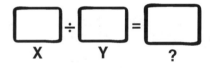
$$X \div Y = ?$$

Make your own **MATH MYSTERIES**

PAGE 69

DIVISION/Fractions

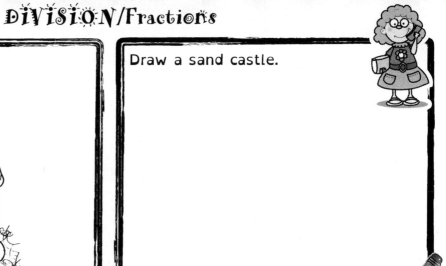

Draw a sand castle.

Pick 1 number

Bella, Rosie, and Sammie want to build a sand castle. They need to get X buckets of sand to build the castle. If they each try to get the same number of buckets, how many buckets will they each need to get?

☐ X

Answer

☐ ÷ 3 = ☐
X   ?

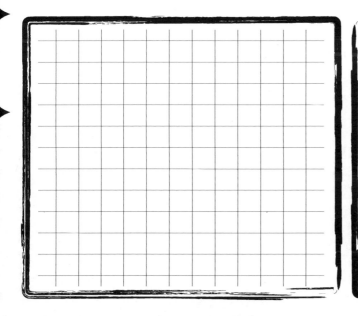

Make your own MATH MYSTERIES PAGE 71

DIVISION/Fractions

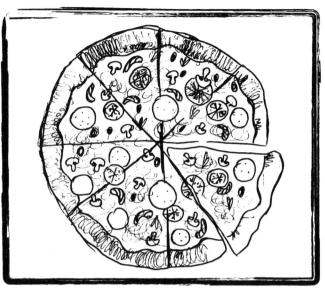

Draw a picture of your pizza party.

Pick 2 numbers
The first number should be 1-5

☐ ☐
X Y

You are having a pizza party! You have X guests coming to the party. The pizza has Y slices. If everyone gets the same amount of pizza, how much does each person get?

Answer

☐ ÷ ☐ = ☐
Y   X   ?

PAGE 72

Make your own **MATH MYSTERIES**

DIVISION/Fractions

Draw a pie and color it to show how much each person gets.

Pick 1 number

☐
X

Arya, Jackson, Antoni, and Lana are sharing a pie evenly. The pie has X pieces. How much of the pie will Lana get?

Answer

☐ ÷ 4 = ☐
X   ?

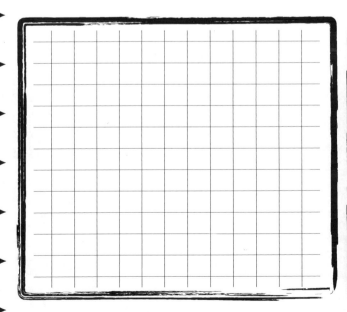

Make your own **MATH MYSTERIES**

DIVISION/Fractions

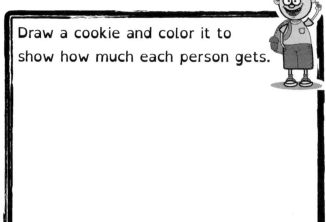

Draw a cookie and color it to show how much each person gets.

Pick 1 number

☐
X

Bonnie, Naomi, and Sasha are splitting a cookie. There are X pieces of cookie. Bonnie gets half of the cookie. Naomi and Sasha share the rest evenly. How much of the cookie did Sasha get?

Answer

$(\square_X - (\square_X \div 2)) \div 2 = \square_?$

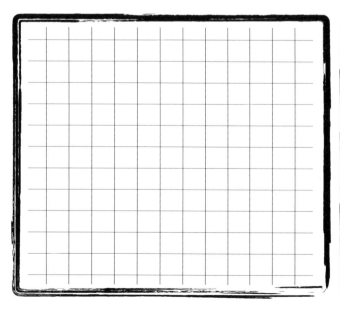

Make your own MATH MYSTERIES

DIVISION/Fractions

Draw some monkeys in their habitat.

Pick an even number

☐ X

You visit the zoo and go to see the monkeys. There are X monkeys in the exhibit. If 1/2 of the monkeys are boys, how many girl monkeys are there?

Answer

☐ ÷ 2 = ☐
X   ?

_____
_____

Page 76   Make your own MATH MYSTERIES

DIVISION/Fractions

Draw your friends on the swings.

Pick 1 number

□ X

You and X friends are at the park and want to play on the swings. There are only four swings for you to share. You get into groups so that each swing has as close to the same number of people as possible. How many of you will ride on each swing? (Be careful, Some swings may have an extra person)

Answer

_____

$(\square_X + 1) \div 4 = \square_?$

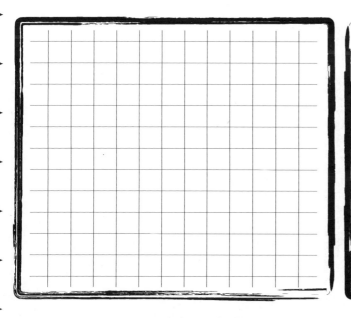

Make your own **MATH MYSTERIES**

DIVISION/Fractions

Draw a pie that shows how much you get to eat.

Pick 1 number

You baked a pie. You want to share it with X friends. If everyone gets two slices, how much of the whole pie do you get? Show it as a fraction. It should look like this. :

$$2 \div ((\boxed{}_X + 1) \times 2)$$

Answer

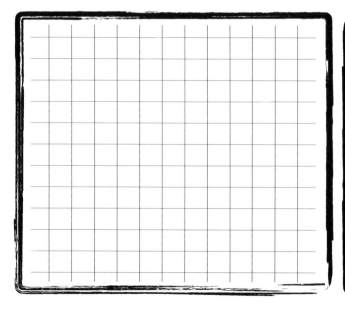

Make your own MATH MYSTERIES

Copyright Information

Thinking Tree Books are created for home and family and individual use only.

You may make copies of these materials for only the children in your household, or purchase books individually for your classroom. All other uses of this material may be used only by permission granted in writing by The Thinking Tree LLC. It is a violation of copyright law to distribute the electronic files or make copies for your friends, associates, or students without our permission.

Contact Us:

The Thinking Tree LLC

317.622.8852 PHONE (Dial +1 outside of the USA) 267.712.7889 FAX

FunSchoolingBooks.com

Made in the USA
Lexington, KY
05 March 2018